Swarm Control

ISBN: 978-1-912271-59-7

Published by Northern Bee Books © 2020

Northern Bee Books, Scout Bottom Farm
Mytholmroyd, Hebden Bridge, HX7 5JS (UK)

www.northernbeebooks.co.uk

Tel: 01422 882751

This booklet is based on an article that first appeared in BBKA News, April 2020.

Photos by author unless stated otherwise.

Design by SiPat.co.uk

Swarm Control

Richard Ball

About the author

Richard Ball

Richard started beekeeping in 1983 helping a small bee farmer and then with his own bees in 1984. After retirement from a previous career in 1996 he was appointed as a Seasonal Bee Inspector for the National Bee Unit with responsibility for Surrey and Sussex. Then in 1999 as Regional Bee Inspector for the South West of England, moving to Colaton Raleigh in Devon and in April 2006 to National Bee Inspector having responsibility for the Bee Inspection Service for England and Wales. He retired from this post in 2009 continuing within the Bee Unit as an Extension Officer until finally retiring at the end of 2011.

Following the discovery of pyrethroid resistant varroa mites in Devon during 2001 he has tried to raise awareness of the importance of Integrated Pest Management for mite control along with other bee disease issues.

Richard still keeps bees and is Chairman of the Devon Apicultural Research Group. His current bee interests include identifying pollen collected by bees and the effects of climate change on the cycle of honeybees in SW England. He has lectured throughout the United Kingdom and also in the Republic of Ireland, Malta and Romania.

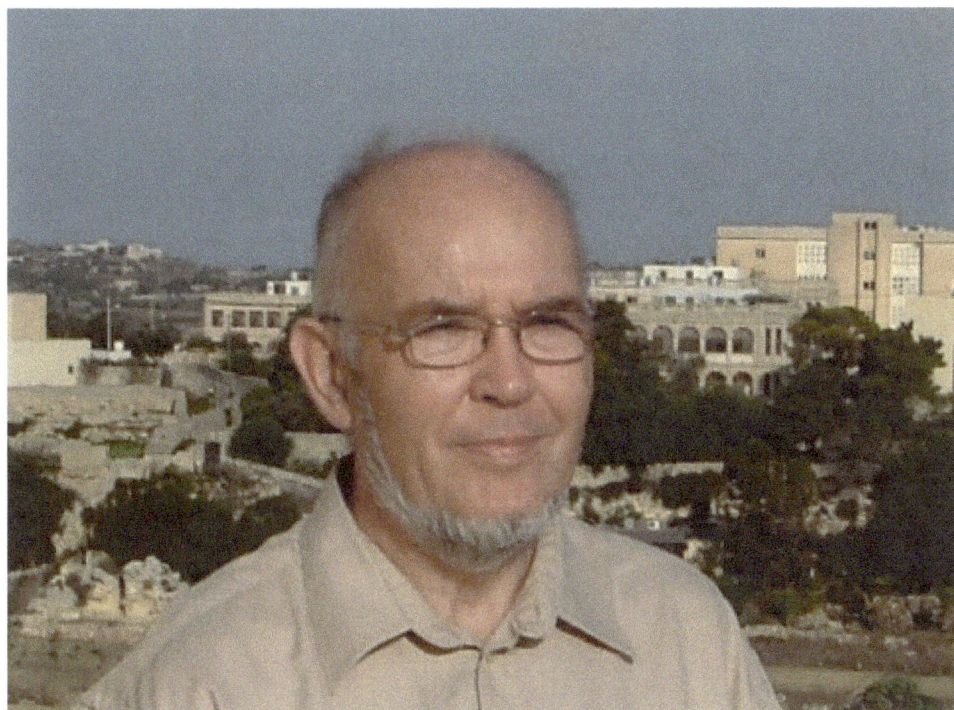

Contents

Swarms

Swarms are a natural phenomenon and are the reproductive cycle of many supraorganisms. These are a group of synergetically interacting organisms of the same species, generally being eusocial insects where individuals do specific tasks and cannot exist on their own for an extended period of time. The most common are ants and honeybees. Their reproductive phase should not be looked upon as the day to day brood rearing but as the supraorganism dividing by means of a swarm to form another one. In the U.K. honeybees swarm between April and August, but it is not unknown for a swarm to occur outside this period particularly in cities where there is more shelter and warmth.

In the case of honeybees the cycle of events starts with the construction of queen cups which tend to be toward the centre of the brood nest and around its periphery. The queen lays eggs in these cups over a period of days and when hatched are fed royal jelly lavishly until they are sealed over on the 9th or 10th day after egg laying. During this period the queen will reduce egg laying resulting in her abdomen becoming smaller and making her more capable of flying. If weather conditions are favourable after the sealing of queen cells excitement levels in the colony rise, bees fill their honey sacs and start to pour out of the entrance, firstly crowding around the entrance and then taking to the air and forming a cloud which thickens. The queen is often one of the last

Figure 1: A swarm of bees on a bush

to leave. This cloud of bees usually settles on a nearby projection, such as a tree or bush, and then entwine themselves into a compact cluster. (*Figure 1*). Scout bees are by then looking for a suitable cavity in which to set up a new nest. The cluster will remain intact for a few hours or even days before taking to the air and flying to their new nest site. This type of swarm is known as a 'prime swarm'. Rarely, before the swarm emerges, the scout bees will have already found a new nest site in which case they omit the cluster phase.

In the original colony, known as the 'parent' the remaining bees redistribute their labour and the queen cells are left to mature. After 5 or 6 days from sealing the first virgin queen will emerge. One of two things then happen. She will either be allowed to destroy her unborn sister queens by stinging the pupae to death through the weaker side walls of the queen cells and take over the colony, or she is stopped from doing this and the colony swarms again headed by a virgin queen. This type of swarm is known as a 'cast'. Of course, the eggs that were originally laid in the queen cells would have been laid over a number of days, so the resulting queens will mature over a similar number of days. Therefore depending on the stock and weather a number of casts may emerge over a number of days. Sometimes two virgins emerge into the colony and they may fight each other until one is killed and the other takes over the colony or they may even leave and fly together with the same swarm subsequently fighting each other to see who will head the new colony.

If swarms are left unchecked honey production will suffer and the parent stock may become a liability for the beekeeper. If collected during the early part of the season prime swarms usually develop into vigorous and productive units. Swarms are seen by many as a public nuisance, so beekeepers need to manage their bees to prevent this nuisance as well as maintaining honey production. Similarly, if your bees swarm you should collect and hive it. Understanding the natural history of swarming is the first step in being able to control them.

Finding queens

The procedures referred to in this booklet require you to find the queen. (*Figure 2*).

This can be difficult for some beekeepers but should become easier with experience. Various techniques can help to find her. The first is to look for her during a nectar flow as more bees will be foraging - so there are less at home to confuse the situation. In extremis the hive can be moved a few feet

Figure 2: A queen heading for dark side of the comb via the bottom bars. Her white mark has almost worn off.

away so the flying bees cannot find their way home and will be seen searching around the original site. This is particularly useful with a defensive colony. The use of cover clothes keeps the combs dark and maintains a little more normality. Established queens have distinctly longer abdomens, lighter coloured legs and tend to swagger around as if they own the colony. Virgin queens have distinctly shorter triangular abdomens and run around like demented beings.

To find the queen in a single brood box colony use the minimum amount of smoke to subdue the colony, otherwise the queen may run to a part of the hive where she will be difficult to find. This is particularly the case with hives using long lugged frames as there is often a dark cavity at the bottom of the brood box where it meets the floor below the frame ends. If possible, examine the colony with the sun at your back. Remove the supers and queen excluder being sure to examine the underside of the excluder for her. When colonies do not have supers on them but just have a crown board over the brood box it is common to find the queen on this board having been driven up by smoke. Remove the end comb and examine it carefully for the queen. If she is not present place the frame in an empty nucleus box or travelling box. If these are not available place the comb adjacent to the hive entrance. If eggs are present do not let the sun shine into the comb as it will desiccate the eggs.

The next frame is then removed but before examining it look down the exposed side of the next comb, and the floor and side walls, where she might be spotted. Then examine the removed comb for the queen. Queens live in a dark environment so when examining the comb look around the edges first as she will often be seen walking around the sides of the comb to go to the dark side. For this reason, when examining a comb turn it around and look at the dark side first. It is also more probable to find the queen on a comb containing eggs, but this is not necessarily the case due to the disturbance caused during the examination. If bees cluster on the frame it is easy to disburse them by gently blowing on them or touching them gently with the flat of your fingers. When you have examined that comb replace it into the brood box adjacent to the end wall and continue to examine the remaining frames. Make sure you replace the frames in their original orientation and position. If you have examined all the frames and not found the queen work your way back across the box examining each frame in turn. I find that if I have not found her in two sweeps, I will not find her. I would never exceed three sweeps.

If a queen is particularly difficult to find get an empty brood box and stand it on an upturned roof at the back of the hive you are searching. Place the first two frames into the new box keeping them together but about five centimetres from the end wall. Then take out the next two frames and again keeping them together place them five centimetres from the first pair of combs. Repeat again and then separate the remaining combs in the original brood box so that they are spread out in the same way. The light will be shining into the hive so the dark areas that the queen will prefer are those between the paired combs. After a couple of minutes remove a pair of combs and open them like a book and you will have a good chance of spotting her.

With double brood chambers a slightly different technique is used. Smoke the colony quite heavily at the entrance and wait for a couple of minutes before quickly removing the supers and then the upper brood box standing it on an upturned roof. The smoking will drive the queen upwards, so she is likely to be in the upper brood box. Because of this make sure you examine the underside of the queen excluder carefully. Examine the combs as described previously.

If the queen is seen to take flight during this manipulation it is best to leave the hive open with the top bars exposed for a short while, so that the scent of the colony attracts her back.

Marking queens

Having a marked queen is a great help in finding her. This is done by marking her with a coloured spot on the thorax. Different coloured spots can be used to remind you of her age. The accepted standard is:

Years ending in	Colour
0 or 5	Blue
1 or 6	White or grey
2 or 7	Yellow
3 or 8	Red
4 or 9	Green

Coloured marker pens are available from bee equipment suppliers but before using them on a queen test them out on the back of your hand in case the ink floods out. Also available are coloured numbered disks that can be cemented on the thorax which are very effective. Many beekeepers use enamel paint, but it must be remembered that there is a risk in using untested materials. Queens should not be marked early in the season and care is required in handling them. Beginners, people with poor eyesight, or those with unsteady hands, are best advised to use one of the various holding down cages such as the Baldock cage. The usual method is to pick up the queen by her wings using the thumb and forefinger of your right hand and then transferring her to the other hand holding her thorax gently between thumb and forefinger from behind her. A dab of chosen marker is then placed on her thorax before releasing her back into the colony. Left-handed beekeepers need to use the opposite hands. If you are worried about injuring your queen practise on some drones.

There is a risk of 'queen balling' which means that a number of bees cluster around her and whilst not directly trying to kill her by stinging they can suffocate or starve her to death. There are two main causes for this, disturbance early in the season or by her acquiring a strange scent, which is often from beekeepers using highly scented soap or handling chemicals. If you see it happening smoking the ball may help to stop it.

Swarming impulses

Apart from the natural desire to multiply there are many factors that influence a swarming impulse in honeybees. The usually quoted ones are congestion in the brood nest, using a small hive and the age or fecundity of the queen. However, there are many more, such as weather, locality, management practised, intermittent nectar flows, race or strain of bee, etc. This makes it difficult to advise other beekeepers and perhaps explains why if you ask two beekeepers for advice you get more than two different answers! With the effects of climate change these impulses are likely to fluctuate far more than in the past, so it is important to monitor and take management measures to reduce this impulse. In another 50 years or so we may well see a total change in U.K. honeybee behaviour with no brood in mid-summer and two brood peaks in spring and autumn. Needless to say, this will make significant changes to swarm impulse and our management practices.

Management to prevent swarming

At our early spring inspection, in March or early April, beekeepers check for many things including the presence of sufficient stores, but this inspection should also check as to whether the colony is congested with stores. This can be a particular problem in smaller hives such as Modified Nationals or WBC's, or if blanket winter feeding has occurred. However, if you have a store of brood frames containing empty clean drawn comb it is a simple matter to take out end frames full of excess stores or damaged ones if you placed them there in the previous season and then insert new drawn comb next to the brood nest. At this early time, it is not practical to use foundation as without a nectar flow the bees will not draw it efficiently, so it becomes dirty and holes are chewed in it. By the time there is an adequate nectar flow a good queen will have laid brood in end combs. These drawn combs are easily produced for use in smaller hives by using brood boxes containing frames of foundation as supers and extracting the honey (*Figure 3*). Of course, you need an extractor that will accept frames of that size. A brood box used as a super is heavy when full so if this is too much for you two five frame nucleus boxes side by side is an alternative. These extracted frames of comb can be stored overwinter ready for use in the following spring. Some beekeepers, such as the late Fred Gale of Gales Honey fame, only use brood boxes so that they always have a plentiful supply of clean brood comb.

Figure 3: A brood box of foundation used as a super.

As the season progresses, we need to ensure there is sufficient super space and the rule of thumb is to add a super when the previous one is about two thirds full. A useful sign is that when needed bees will be building comb between the crown board and the frames below. Adding supers at an earlier stage causes 'chimneying' where the end frames in the supers are left unused.

Requeening with a young fecund queen can reduce the swarming impulse.

It appears that varroa infestations increase the swarming impulse so efficient varroa control is an important part of swarm control.

In my view there is no doubt that the smaller hives, without good management practices, are generally too small for most bees used in the U.K. After all a reasonably good queen can lay 2,000 eggs a day when the colony is building up, but a Modified National Hive theoretically only has space for laying a maximum of 1,247 eggs per day. This creates a classic swarm impulse caused by congestion. A 'Demaree' can be performed to reduce this impulse. This procedure gives the queen more laying space before the swarm impulse arises so should not be used if there are eggs or larvae in queen cells.

Demaree

1. Examine the colony and find the queen. Make sure she is on a frame of brood and place her in another clean brood box filled with frames of clean empty drawn comb.

2. Reassemble the hive as follows:
   Floor – Brood box containing queen and drawn comb – Queen excluder – Any supers – Original brood box containing frames of brood – Crown board – Roof

3. Because the brood is at a distance from the queen there is insufficient queen pheromone present and a swarm impulse occurs in this box. Therefore, after seven days it is essential to examine these frames and destroy any emergency queen cells present (*Figure 4*).

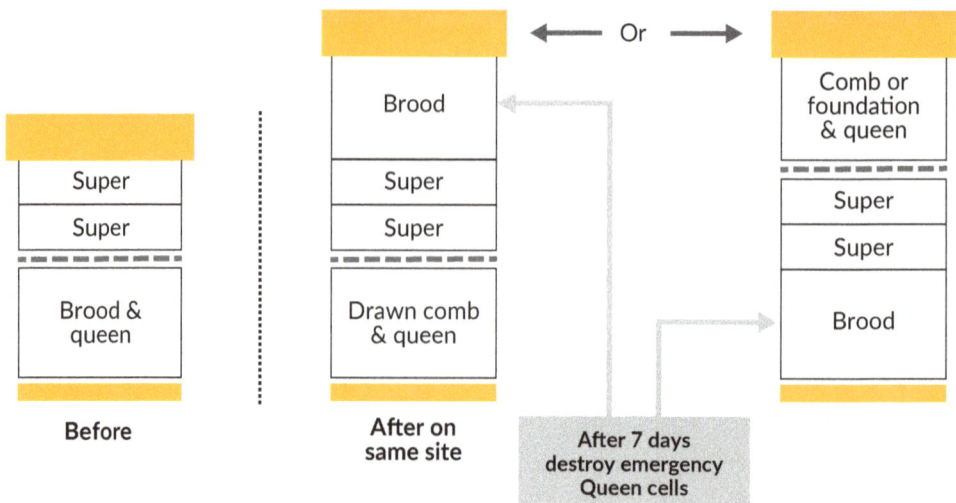

Before — After on same site — Or — After 7 days destroy emergency Queen cells

Figure 4: Emergency queen cells. Note the hooked nose appearance. They are easily overlooked.

With this system it is not possible to use frames of foundation, only drawn comb. This is because with the brood on top the young bees will tend to remain there and the foundation will not be drawn out efficiently and become dirty and spoilt.

A variation of Demaree

A variation of Demaree is to reassemble the hive as follows:

Floor – Original box containing frames of brood – Any supers - Queen excluder – Brood box containing queen and drawn comb – Crown Board – Roof

This variation is useful if you wish to change old brood comb because it is possible to use frames of foundation in the top box. Thus when the brood has hatched in the lower box the bees tend to move any suitable unused stores upwards as the combs are abandoned. Once this happens the old brood box and combs can be removed, and the hive re-assembled in the conventional manner.

Swarm control

In normal bee management colonies are examined every 9 days and if no queen eggs, larvae or sealed cells are present then the colony can be closed up and there is no risk of a swarm emerging before the next 9-day examination. This is because colonies do not usually swarm until queen cells are sealed. If clipped queens are used this inspection period can be increased to 14 days, but there is a risk of a swarm during the last five days of this period. This swarm will emerge but because the queens wing is clipped she will be unable to fly properly and fall to the ground close to the hive (*Figure 5*). A few bees will cluster round her up to the size of a golf ball and she will perish unless recovered by the beekeeper. The swarm meanwhile will settle as usual and after about thirty minutes will realise they do not have a queen and return en-mass to the hive (*Figure 6*). If management does not occur within the 14 days then a virgin queen will emerge, followed by a swarm. In my experience these swarms are particularly large! An examination at the 14-day period will reveal queen cells, no eggs or larvae and an intact colony. One cell that is about to hatch is left and the others destroyed. Remember clipping queens' wings does not prevent swarming, it only delays a swarm decamping.

If you wish to clip a queen's wing to prevent her flying catch the queen between thumb and forefinger holding her gently by the thorax as described before. Using a sharp pair of small scissors cut the tip off one large wing. Reducing this to the size of the small wing is sufficient to put her off balance when flying. Never clip the wing of a virgin queen!

Figure 5: A queen with a clipped wing in grass adjacent to her hive being unable to fly when swarming. She is marked with a numbered disk which has worn so the number is not visible.

Figure 6: A swarm that has returned to a hive having lost their queen because her wing was clipped.

When eggs or larvae in queen cells are found it is common practise to destroy them and this is probably o.k. on first discovery but remember queen cells should not be destroyed unless you are confident that the queen is present or at least that eggs are present. Repeated destruction usually ends in disaster when a cell is missed and, in any case, colonies treated this way become disillusioned and fail to perform.

When destroying queen cells make sure that the eggs and larvae are killed or removed as the colony can repair damaged cells containing a viable egg or larva. A common mistake made by inexperienced beekeepers is to destroy all the queen cells when a swarm has already emerged thus leaving the colony queen less and with no means of making a replacement. Sometimes a hatched queen cell will be found (*Figure 7*) when it is easy to assume that a virgin queen is present. However, if a cast has emerged you will force the colony into a queen-less state by destroying all the remaining queen cells. So unless a virgin queen has been seen the action to take is to find a queen cell that is about to hatch, these have a dark ring around the base, and pop the end cap off with the corner of a hive tool. If you are lucky you may see a virgin queen chewing around this dark line to release herself, a good one to choose. You can then observe the virgin queen run out onto the comb. The other cells can then be destroyed in the knowledge that the colony has at least one virgin queen and no more can hatch. Of course, if the original hatched cell virgin queen is still present you may lose a caste swarm, but you don't lose the colony through repeated swarming.

Artificial swarming

It is better to follow the bees and manage their natural desire. There are many techniques to do this and, in this booklet, we will consider the principal ones. Most rely on the fact that flying bees return to the site of their hive, not the hive itself. To demonstrate this, move one of your hives say 70 cms. to one side and observe the returning forager bees. They will fly to the original position become confused and circle around for some time before finding their own hive. If another hive is placed on the original location the foraging bees use it as their own. A process of deliberately drifting bees in this way is the basis of Artificial Swarms.

The main techniques are called Pagden, Heddon and Snelgrove, which many beekeepers develop to suit their own needs and have also been developed to control varroa. Here we will concentrate on and fully describe Pagden and Heddon.

Figure 7: A hatched queen cell.

When performing these techniques, it is essential that the bees are flying freely on a nice warm, dry day.

Pagden

When queen eggs or larvae are found, a new hive filled with foundation or clean drawn comb is placed by the hive being examined. Then:

1.   The queen is found and if she is on a frame of sealed brood the frame is placed in the centre of the new hive. If the queen is on open brood it easy to hold a frame of sealed brood above and against it, then by gently smoking she can be moved up onto it.

2.   The original hive, known as the 'parent' is then moved to another site in the apiary.

3.   The new hive, known as the 'swarm' with the queen, one frame of sealed brood and frames of foundation or drawn comb is placed on the original site. The queen excluder and any supers are placed on this hive and it is closed up. Bees returning from foraging join up with the bees in the supers and the queen.

4.   The parent colony is then fully examined and one queen cell with a healthy larva on an abundant bed of royal jelly is selected (*Figure 8*). It is best to select one which will not be damaged by subsequent examinations. Mark the frame top bar above its location so you know which you selected. A drawing pin is useful for this (*Figure 9*). Carefully examine the rest of the frame and destroy any other queen cells. Do not shake the bees off this comb due to the risk of dislodging your chosen larva. Then examine every other brood frame and destroy any queen cells present on these frames. It is best to shake off the bees so as not to overlook any cells. To shake bees off a brood comb, lift the comb partially out of the brood box, hook your finger round the frame lugs and return the frame rapidly into the box stopping suddenly about 3 cm. from the top of the box. Most of the bees will fall unharmed to the hive floor.

5.   Check they have sufficient stores until the next examination and when completed close up the hive.

6.   After seven days examine the parent colony and destroy any emergency queen cells built (*Figure 4*). Be careful not to shake the marked comb or destroy your chosen cell.

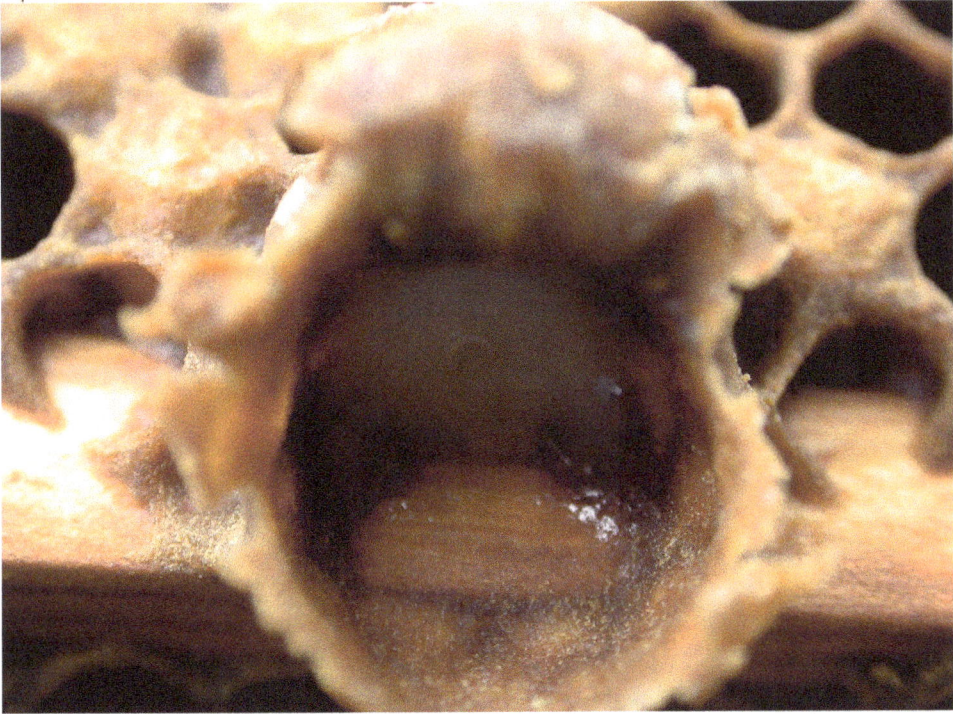

Figure 8: A queen larva on a bed of royal jelly. The edges of the cell have been peeled back for the photograph. They must not be peeled back on a selected cell.

Figure 9: A top bar marked with a drawing pin. This is clearly visible even with bees running over it.

7. Also examine the swarm colony as when a colony is artificially swarmed it does not necessarily settle down in its new home despite the presence of brood. Much depends on the state of advancement of the swarming impulse in the original hive, so you may find developing queen cells! These can be destroyed.

Although Pagden can work well an imbalance between swarm and parent occurs. This is because although the parent stock is depleted of flying bees it soon gathers impetus as the brood hatches out and the new queen rapidly increases egg laying. The swarm colony on the other hand although strong after the manipulation soon dwindles. You may wish to consider this relative to late nectar flows.

The question of using foundation or drawn comb in the swarm colony remains. Generally, there is no choice as the beekeeper doesn't have clean drawn comb. If the swarm is created during or immediately before a good nectar flow it is best to use foundation as the bees will be forced to store nectar in the supers and good comb will be drawn in the brood box for the queen to lay eggs in. However, if the nectar flow is poor or non-existent foundation tends to become dirty, nibbled and spoilt so then clean drawn comb is a better choice. It also gives the queen comb to lay eggs in. The danger with using drawn comb is that if a nectar flow comes on then the combs become clogged with stores. To make the right choice requires a good knowledge of local circumstances and an assessment of forthcoming weather.

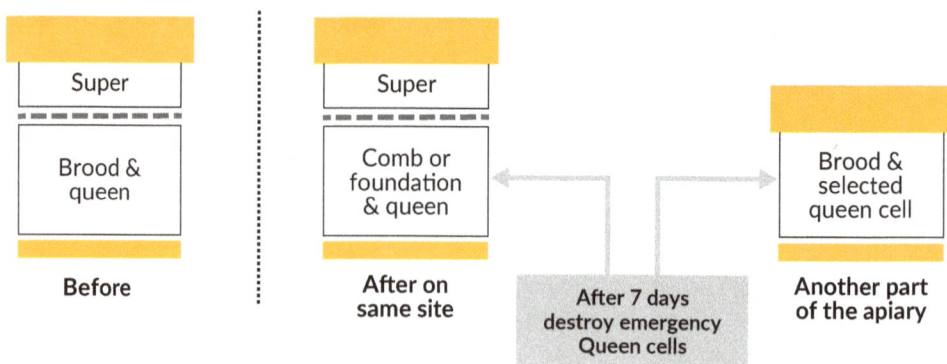

Super		Super			Brood & selected queen cell
Brood & queen		Comb or foundation & queen			
Before		**After on same site**	After 7 days destroy emergency Queen cells		**Another part of the apiary**

Heddon

This is a refinement of Pagden and creates a more balanced split.

1.   The artificial swarm is made up in exactly the same way as in Pagden. The queen is found and if she is on a frame of sealed brood the frame is placed in the centre of the new hive. If the queen is on open brood it easy to hold a frame of sealed brood above and against it, then by gently smoking she can be moved up onto it.

2.   The parent colony is moved to one side and the new hive, known as the 'swarm' with the queen, one frame of sealed brood and frames of foundation or drawn comb is placed on the original site. The queen excluder and any supers are placed on this hive and it is closed up. Bees returning from foraging join up with the bees in the supers and the queen.

3.   The parent colony is now placed next to the swarm colony with the entrance facing away and at right angles to the original entrance (*Figure 10*).

Figure 10: Manipulation for Heddon result in the swarm and parent being located like this. It may require some planning with portable hive stands if you are using twin hive stands like this one.

4. After 2 days the parent is moved to the other side of the swarm with the entrance facing away so is turned through 180°.

5. After a further 2 days the parent is moved to another part of the apiary.

The purpose of these movements is to drift bees off from the parent to the swarm to strengthen it and to weaken the parent, so it is unlikely to throw a caste swarm saving the need to examine the parent colony for queen cells.

Heddon – day 1

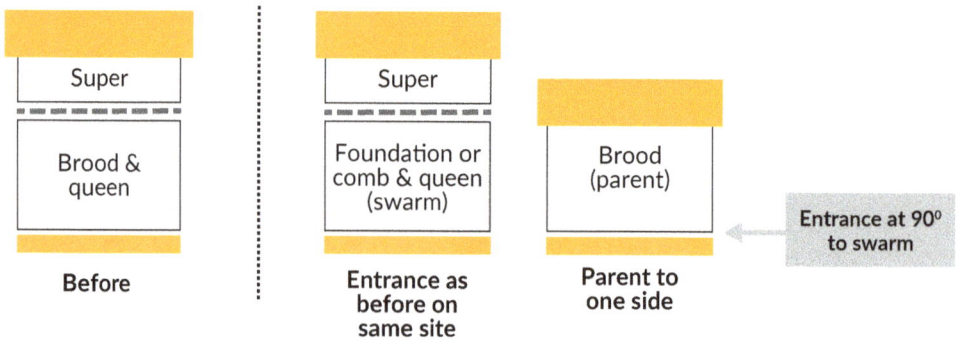

Heddon – day 3

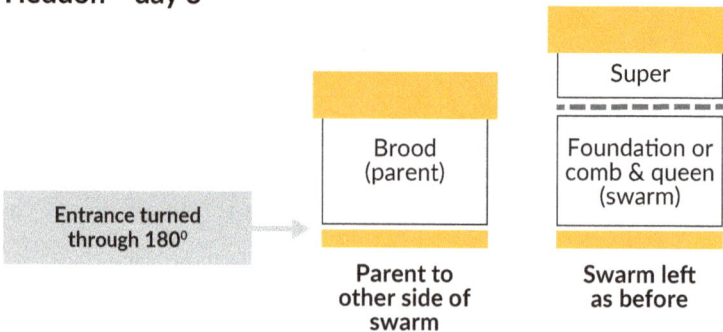

Heddon – day 5

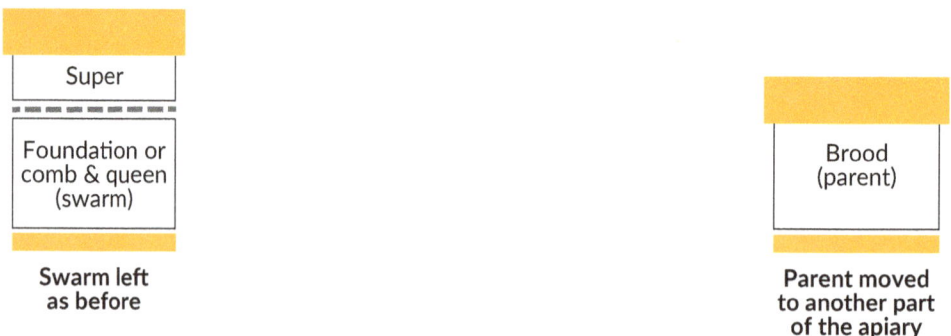

Both these methods rely on drifting bees from side to side and require an extra complete hive. By drifting bees up and down using the Snelgrove method a saving of a floor and roof can be made. However, this method is more suited to a single walled hive as with double walled hives an entrance tube, known as a 'Shepherd tube', needs to be created through the void between the inner and outer walls for the upper parent colony.

Snelgrove

1. The artificial swarm is made up in exactly the same way as in Pagden. See steps 1 & 3 in Pagden above.

2. The new hive, known as the 'swarm' with the queen, one frame of sealed brood and frames of foundation or drawn comb is placed on the original site. The queen excluder and any supers are placed on this hive. A Snelgrove board is then placed on it and the parent colony placed on top (*Figures 11 and 12*).

Figure 11: An original Snelgrove board made by L.E.Snelgrove. Note that it has slides rather than fold out doors. Photo Ken Edwards.

Figure 12: A Snelgrove board in use. The top door is open
and the lower door can be seen. Photo: Jan Stuart.

3.   This board has a mesh centre permitting colony odours to be exchanged between the parent and swarm, and by manipulating a number of doors bees can be drifted between the upper and lower boxes to replicate the `Heddon' method. These doors are placed on three sides of the crown board being on the upper and lower side of the board. It is easy therefore to drift bees between the swarm and parent.

This method works well but many beekeepers find it fiddly and tend to do it once and move on to something they prefer. I would recommend reading L. E. Snelgroves book 'Swarming – Its control and prevention' for full directions.

Modifying Snelgrove

Beekeepers often modify Snelgroves method using standard equipment and it is particularly suited to smaller hives. A standard crown board is adapted so that on the top surface a section is cut out of the bead on one side and by using a small pin or screw converted to a fold out entrance (*Figure 13*). An artificial swarm is created in exactly the same way as for Pagden and when reassembled the crown board is put on, the feed holes completely closed off with a piece of plywood, and the parent colony is placed on top with the fold out entrance open and facing to the rear. Bees can be bled off as in Heddon by turning the parent colony and crown board through 180° over two consecutive two day periods, but the use of a Snelgrove board would be more efficient and save labour.

The disadvantages with these methods are that bees fly at head height and brood boxes being high up are difficult and heavy to handle when examining the colony. However, they do save equipment and space.

Figure 13: A modified crown board with a hinged entrance block. The fold out does not have to be this long.

Can't find the queen!

All the methods described above require the beekeeper to find the queen, but I have found that many beekeepers have difficulty or just can't find them resulting in swarm control being abandoned. So, what do you do if you can't find her? The first thing to say is that if there are sealed queen cells then you are probably too late, and a swarm will have emerged. The system below relies on the same principals of drifting bees as in conventional swarm control.

1. When queen eggs or larvae are found the hive is moved to another part of the apiary, being at least 1.1 metres (four feet) away. This is the 'Parent'. As the bees bleed off to go to the hive at stage '2' this will have the benefit of making the queen easier to find.

2. A new hive, filled with foundation or clean drawn comb, is placed on the original site. This becomes the 'Swarm'.

3. Examine the 'Parent', and if you can't find the queen remove a frame of brood with eggs and young brood present. Make sure it has no queen cells on it. Place this frame in the centre of the 'Swarm', it does not matter if it has bees on it.

4. Place the queen excluder on the 'Swarm' together with any supers and close up with the crown board and roof.

5. Replace the comb you removed from the original hive and close up with a crown board and roof.

The flying bees will go to their original site where the swarm hive is. The young bees and brood stay in their original hive at its new location. This effectively divides the colony in half and will in most cases remove the impulse to swarm. The queen will be in one of the hives.

After seven days:

1. Examine the Swarm. If there are no queen cells and you can see worker eggs then this is where the queen is. If there are emergency queen cells and no worker eggs then select one queen cell and destroy the others.

2. In the original, parent, hive the bees should have decided that if the old queen is there they do not wish to swarm anymore and will break down the queen cells. If the queen is not there, they will bring on the

queen cells and because of the reduced number of worker bees will probably only permit one to hatch.

It is not a guaranteed system but is often effective. However, it is best to develop, and hone queen finding skills and carry out more effective methods of swarm control.

Using an artificial swarm for varroa control

In the 1980s as varroa swept across Western Europe a lot of varroa research was carried out. It was clear to many that just reaching for the chemical or medicine pot for control was not good for bees and wasn't the only answer. A lot of work was put into developing bio-technique (management) control. The principal behind most of these is that if a colony is made brood less all the varroa mites will be on adult bees, the so called phoretic phase. If a frame of open brood is presented all the mature varroa mites enter this brood to carry out their reproductive phase. Therefore when this frame of brood is capped over it can be removed and destroyed along with the mites that it contains. This technique will remove about 90% of the varroa mites in a hive. If two frames of brood are used this is increased to 95% of mites. This efficacy is greater than most medicinal treatments currently available.

Researchers in Germany and elsewhere developed this principal into queen trapping and the use of artificial swarms. The biggest problem with these techniques is that as they are performed early or mid-season so there is a risk of reinvasion from collapsing and heavily infested colonies nearby. This is where, if you unlucky enough to have neighbouring colonies with uncontrolled varroa populations that are collapsing, the remaining able-bodied bees abandon the hives and find a home in other nearby hives. These bees will carry a significant number of phoretic varroa mites and will re-infest the controlled colony.

Many beekeepers worry about the loss of bee brood with systems based on this principal but when timed in regard to local nectar flows and conditions it can increase honey yields. This is because more bees will go foraging when there is no brood present. However, the colony will need to be queen right in time to recover its full strength before late summer. In my South of England environment I find that the end of May and into June are most effective for these procedures.

Artificial swarm to control varroa:

1. The queen is found and if she is on a frame of sealed brood the frame is placed in the centre of a new hive. If the queen is on open brood it is easy to hold a frame of sealed brood above and against it, then by gently smoking she can be moved up onto it.

2. The original hive, known as the 'parent' is then moved to another site in the apiary at least 4 metres away.

3. The new hive, known as the 'swarm' with the queen, one frame of sealed brood and frames of foundation or drawn comb is placed on the original site. The hive is closed up with a new crown board and roof. As this hive has no stores it will need feeding unless there is a very good nectar flow. It is best to use a contact feeder. Bees returning from foraging join up with this colony. The efficacy can be slightly improved by not using the frame of brood and just placing the queen in the new box. However you must place a queen excluder beneath the brood box to prevent the queen and bees absconding.

4. The parent colony is then fully examined and one queen cell with a healthy larva on a bed of royal jelly is selected. This cell is caged so that the emerging virgin can't leave the hive to mate. It is best to select a cell in a convenient position and mark the comb with a drawing pin. Carefully examine the rest of the frame and destroy any other queen cells. Then examine every other brood frame and destroy any queen cells present on these frames. It is best to shake off the bees so as not to overlook any.

5. The queen excluder and any supers are placed on this parent hive and it is closed up.

6. After seven days examine the parent colony and destroy any emergency queen cells built. Be careful not to shake the marked comb which may damage the selected cell or dislodge the cage.

7. After a further 17 days the parent colony will be brood-less, so two frames of open brood are taken from the swarm colony, being careful to shake off any bees, and placed in the parent colony.

8. When these frames are sealed over, 9-11 days later, they can be removed and destroyed. The virgin queen should then be culled and the colony requeened with a new vigorous queen.

There is no need to bait the swarm colony because, unless a colony is collapsing or heavily infested, flying bees do not carry varroa mites. The so called phoretic varroa mites on adult bees feed on the young bees they emerged from the brood cells with until they are mature and ready to reproduce. In a natural swarm, rather than one generated by drifting foragers, the bees are a cross section of ages so a collected swarm will have a substantial number of young bees and may carry a varroa mite burden.

The virgin queen needs to be culled because she will be too old to mate satisfactorily so if left the queen will probably be a drone layer.

Instead of destroying the frames of sealed brood it is possible to uncap them and wash the pre-pupae, pupae and varroa out of the comb with warm water. The comb can then be reused. Personally, I find this rather labour intensive so prefer to destroy it.

Artificial swarm – day 1

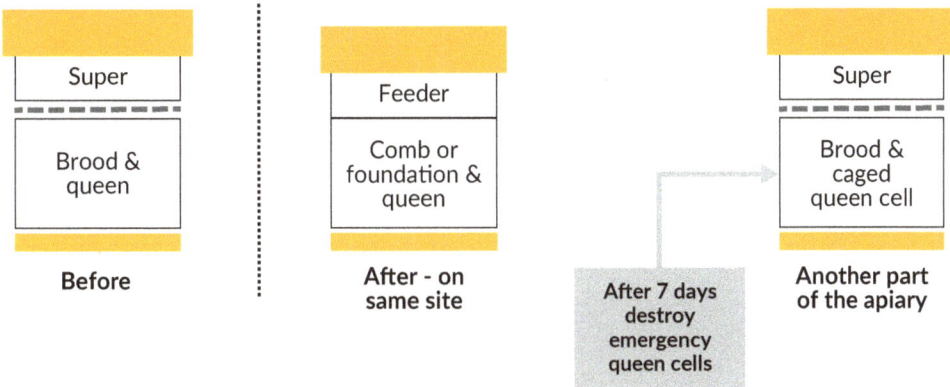

Super		Feeder			Super
Brood & queen		Comb or foundation & queen			Brood & caged queen cell
Before		**After – on same site**	**After 7 days destroy emergency queen cells**		**Another part of the apiary**

Artificial swarm – day 24

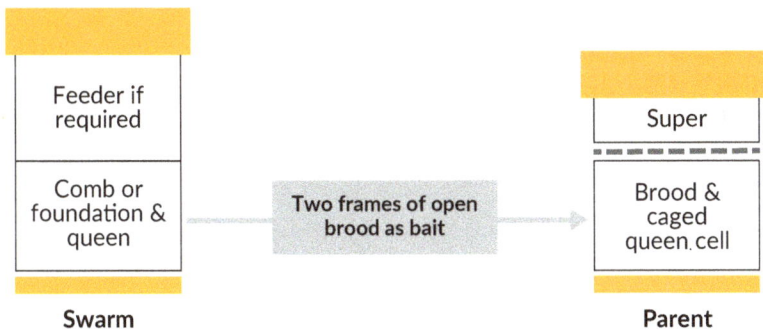

Feeder if required			Super
Comb or foundation & queen	**Two frames of open brood as bait**		Brood & caged queen cell
Swarm			**Parent**

Artificial swarm – day 33

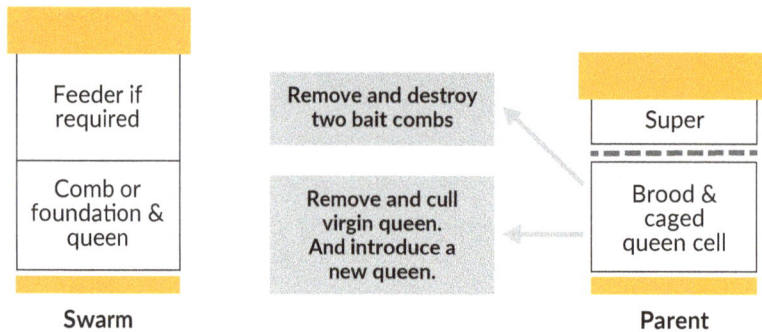

Feeder if required	Remove and destroy two bait combs	Super
Comb or foundation & queen	Remove and cull virgin queen. And introduce a new queen.	Brood & caged queen cell
Swarm		**Parent**

Adapting Pagden to help control varroa

In the U.K. I find that strong colonies which are artificially swarmed tend to go into a brood-less state for some time before the new queen is laying. This is a suitable period to bait the colony with two frames of open brood. My observations of varroa levels in artificial swarms by the conventional 'Pagden' method shows that varroa level remains low in the swarm colony and high in the parent so if you use the conventional 'Pagden' method you can bait the parent colony during its brood-less period with considerable effect.

So carry out steps 1 to 6 in the Pagden section and then:

1. After 24 days examine the Swarm colony and remove one or two frames of open brood making sure there are no bees on them.

2. Examine the Parent, making sure it is brood-less, and place the open brood frames into the brood box. and mark the frames.

3. Replace the removed frames from the Swarm with those removed from the Parent. Make sure no bees are on these combs.

4. After 9-11 days remove the marked frames to destroy them and replace with new framess.

Controlling increase

By controlling swarms with these methods, you start with one colony and end up with two. This may appear great but if nothing is done and you have no losses then the number of hives you have will increase exponentially. If you have 2 hives in year one, and you artificially swarm each colony each season, you will have 912 in ten years; rather expensive in hives and you will be a bee farmer! It is normal practice, unless you require more colonies, to re-unite the hives when the new queen is performing satisfactorily and culling the old queen, however it may be a prudent management practice to run the two colonies through the winter and unite in early spring, thus creating a very strong colony to maximise honey production on spring nectar flows and also hedging your bets against some winter losses.

Uniting

Honeybee colonies have their own particular odour created by the different crops collected and the pheromones within that colony. As a result, mixing bees from different colonies can cause fighting but we may wish to unite bees and colonies as part of our management. During a nectar flow it is possible to shake bees from two or more colonies into an empty brood box and floor mixing them as you go and then placing the combs of brood from the colonies into the new box. Dusting with icing sugars can also help. It is often very effective if three colonies are shaken together. Frames of foundation can be used instead to replace the old brood combs. Though queens can be left to fight it out it is best if one queen is specifically selected and placed in a queen introduction cage in the new united colony.

The newspaper method

If there is no nectar flow, or if you are in any doubt, it is best to use this method. The colonies you wish to unite are moved towards each other remembering the axiom less than 90 cm. (3 feet) or more than 5 kilometres (3 miles). You can move them every day as the bees will have reoriented in 24 hours. The unwanted queen is culled or placed in a nucleus. A sheet of newspaper is then placed over the brood box of the queen less colony. If it is windy it can be held in place using drawing pins or a queen excluder. Using the corner of a hive tool or pen knife a few small slits are made in the paper. The queen right brood box is lifted off its floor and placed over the newspaper. (*Figure 14*) The colonies are able to mix their odours and bees will exchange food through the slits, so by the time the newspaper is eaten away they will have a common smell and will effectively unite. When they have successfully united you will see tiny bits of chewed newspaper piled up outside the hive entrance.

As I wrote earlier swarms are often seen by non-beekeepers as a public nuisance so efficient swarm control is an essential element of responsible beekeeping. The basics of swarm control are outlined in this booklet so it is important to learn and practice how to find queens and develop a system of control that is effective for you.

Figure 14: Uniting hives using newspaper. In this case the queen was removed from a colony with supers resulting in two levels of interaction and newspaper.